PASTURE-RAISED PIG FARMING GUIDE

From Planning And Health Care To Sustainable Practices And Market Strategies

Ethan Harry

Table of Contents

CHAPTER ONE

INTRODUCTION TO PASTURE-RAISED PIG FARMING

Pasture-raised pig farming is an agricultural practice where pigs are raised outdoors on pasture, as opposed to being confined in indoor facilities. This method of farming allows pigs to roam freely, forage, and engage in natural behaviors, which can lead to healthier animals and higher-quality meat.

One of the key aspects of pasture-raised pig farming is the emphasis on animal welfare. Pigs are given ample space to move around, root in the soil, and enjoy fresh air and sunlight. This contrasts sharply with conventional pig farming, where animals are often kept in crowded, indoor environments with limited movement. By allowing pigs to express their natural behaviors, pasture-raised systems can reduce stress and improve overall health.

Another important feature of pasture-raised pig farming is the diet of the animals. Pigs on pasture can eat a diverse diet that includes grasses, roots, insects, and other natural forage. This diet can be supplemented with grains and other feed, but the primary source of nutrition comes from what the pigs find in their environment. This varied diet not only contributes to the pigs' well-being but also enhances the flavor and nutritional quality of the meat.

Pasture-raised pig farming also has environmental benefits. The practice of rotational grazing, where pigs are moved to different sections of pasture regularly, helps maintain soil health and prevent overgrazing. The pigs' natural behavior of rooting and foraging can aerate the soil, promote plant growth, and improve nutrient cycling. This can lead to more sustainable land use and reduce the need for chemical fertilizers.

Farmers who choose pasture-raised methods often find that it aligns with their values of sustainability and ethical animal treatment. However, it does come with challenges. Managing pigs outdoors requires more land and can be labor-intensive. Farmers need to ensure that pigs are protected from predators and that their pastures are maintained properly. Weather conditions can also impact pasture availability and pig health.

Despite these challenges, many consumers are willing to pay a premium for pasture-raised pork due to its perceived health benefits, superior taste, and ethical considerations. As awareness of animal welfare and sustainable farming practices grows, so does the demand for pasture-raised products.

Pasture-raised pig farming is a method that prioritizes animal welfare, environmental sustainability, and high-quality meat production. While it requires careful management and can be more labor-intensive than conventional methods, it offers significant benefits for the animals, the environment, and consumers seeking ethically produced food.

WHAT IS PASTURE-RAISED PIG FARMING?

Pasture-raised pig farming is an agricultural practice that emphasizes allowing pigs to roam freely in a natural, outdoor environment rather than confining them to small, indoor spaces. This method is rooted in principles that prioritize the welfare and natural behavior of the pigs, resulting in healthier animals and higher-quality pork.

In a pasture-raised system, pigs are given ample space to move around and express their natural behaviors. Unlike conventional farming methods, where pigs are often kept in restrictive indoor pens, pasture-raised pigs have access to open fields where they can forage for food. This foraging behavior is integral to their natural diet and lifestyle. Pigs in the wild spend a significant portion of their time rooting in the ground, searching for a variety of foods such as grasses, roots, insects, and other small organisms. By allowing pigs to engage in these activities, pasture-raised farming not only meets their nutritional needs but also supports their psychological well-being.

The diet of pasture-raised pigs is diverse and largely derived from the natural environment. While they primarily graze on pasture, consuming grasses and other plants, this diet is typically supplemented with feed provided by the farmer to ensure balanced nutrition. The supplementary feed often includes grains, legumes, and other nutrient-rich components, tailored to complement the pigs' foraged diet. This combination of natural and supplemental feeding ensures that the pigs receive all the necessary nutrients for optimal growth and health.

One of the key benefits of pasture-raised pig farming is the improvement in animal health and welfare. Pigs that are allowed to live in a more natural environment tend to have stronger immune systems and are less prone to diseases that are common in confined, intensive farming systems. The ability to move freely and exercise reduces stress and promotes physical fitness, which in turn contributes to the overall well-being of the animals. Additionally, the exposure to fresh air and sunlight further enhances their health.

Another significant advantage of pasture-raised pig farming is the positive impact on meat quality. Pigs that are raised on pasture typically produce pork that is richer in flavor and has a better texture compared to meat from conventionally raised pigs. The varied diet and increased activity levels contribute to the development of marbling and fat content in the meat, resulting in a more desirable product for consumers.

Moreover, pasture-raised pig farming can have environmental benefits. Managed correctly, this farming method can promote soil health and biodiversity. The pigs' natural rooting behavior helps to aerate the soil, and their manure acts as a natural fertilizer, enriching the land. Rotational grazing practices, where pigs are moved between different pasture areas, can prevent overgrazing and allow vegetation to recover, supporting sustainable land use.

BENEFITS OF PASTURE-RAISED PIG FARMING

Pasture-raised pig farming offers numerous advantages that span animal welfare, meat quality, environmental sustainability, and economic profitability. This holistic approach to pig farming aligns well with increasing consumer demand for ethically produced, high-quality food products.

Animal Welfare

One of the most significant benefits of pasture-raised pig farming is the improvement in animal welfare. Pigs in conventional farming systems are often confined to small, overcrowded spaces, which can lead to high levels of stress and related health issues. In contrast, pasture-raised pigs have ample space to roam, root, and engage in natural behaviors. This freedom to move and interact with their environment reduces stress levels, leading to healthier, happier animals. The opportunity for pigs to exhibit natural behaviors such as foraging and wallowing not only enhances their physical health but also their psychological well-being.

Healthier Meat

The diet and living conditions of pasture-raised pigs contribute to the production of healthier meat. Pigs that are raised on pasture consume a more varied and natural diet, including grasses, roots, and insects, which enhances the nutritional profile of the meat they produce. Pasture-raised pork is typically higher in omega-3 fatty acids, which are essential for heart health and reducing inflammation. Additionally, this meat contains higher levels of vitamins such as vitamin E and antioxidants, which contribute to overall health benefits for consumers. The combination of a more natural diet and reduced stress results in meat that is not only healthier but often more flavorful than conventionally raised pork.

Environmental Benefits

Well-managed pasture systems provide several environmental benefits that make this farming method more sustainable than conventional practices. Pasture-raised pig farming can significantly improve soil health through natural fertilization from pig manure, which adds essential nutrients back into the soil. This natural fertilization process can reduce the need for chemical fertilizers, which are often associated with environmental pollution. Furthermore, the rooting behavior of pigs helps to aerate the soil, promoting better water infiltration and reducing soil erosion. Pasture systems also enhance biodiversity by creating habitats for various plant and animal species, contributing to a more balanced and resilient ecosystem.

Economic Advantages

From an economic perspective, pasture-raised pig farming can be highly advantageous for farmers. Consumers are increasingly willing to pay a premium for meat that is marketed as being ethically raised and of higher quality. This demand allows farmers to command higher prices for pasture-raised pork, which can improve their profit margins. Additionally, farmers who adopt pasture-raised systems often find that their operational costs are lower in the long run. Reduced reliance on expensive feed and veterinary treatments, along with potential savings on fertilizers due to natural soil enrichment, contribute to overall cost efficiency. Marketing pasture-raised pork also opens up opportunities to attract health-conscious and environmentally aware consumers, further enhancing market potential and profitability.

HISTORICAL CONTEXT AND MODERN PRACTICES

Historically, pig farming was an activity deeply rooted in agrarian lifestyles, where pigs were often raised in open fields or woodlands. These environments allowed pigs to forage freely, engaging in their natural behaviors such as rooting and exploring. This traditional method of raising pigs was widespread before the advent of industrial farming practices. It was not only beneficial for the pigs, who could exhibit their natural behaviors, but also for the ecosystem, as pigs contributed to soil health through their foraging activities.

As the 20th century progressed, agriculture underwent significant changes with the rise of industrial farming. This shift was driven by the need to meet the growing demand for food in an increasingly populous world. Pig farming became industrialized, characterized by the confinement of pigs in indoor environments. These indoor systems were designed to maximize efficiency and production, often at the expense of the animals' natural behaviors and welfare. Pigs were housed in controlled environments where they had limited space and were deprived of opportunities to forage or root in the soil. While these methods increased productivity and allowed for better control over disease and feeding, they also raised concerns about animal welfare and environmental impact.

In recent years, there has been a resurgence of interest in sustainable and humane farming practices, including pasture-raised systems. This renewed interest is driven by a growing awareness of animal welfare, environmental sustainability, and the desire for more ethically produced food. Modern pasture-raised pig farming combines the best of traditional methods with contemporary knowledge about animal husbandry and environmental management. This approach seeks to balance productivity with the welfare of the animals and the health of the land.

Pasture-raised systems allow pigs to roam freely in open pastures, where they can engage in natural behaviors such as rooting, foraging, and socializing with other pigs. These systems often rotate pigs between different pastures to prevent overgrazing and to allow the land to recover, a practice known as rotational grazing. This method not only benefits the pigs but also promotes soil health and biodiversity. By foraging, pigs help to aerate the soil and control weeds, reducing the need for chemical inputs.

Moreover, pasture-raised pig farming is aligned with the principles of regenerative agriculture, which aims to restore and enhance the health of the ecosystem. This approach reduces the reliance on synthetic fertilizers and pesticides, improves soil fertility, and increases carbon sequestration. It also supports a more resilient farming system that can better withstand environmental challenges such as droughts and floods.

The shift towards pasture-raised systems reflects a broader movement in agriculture towards more sustainable and humane practices. Consumers are increasingly demanding transparency in food production and are willing to support farming practices that prioritize animal welfare and environmental stewardship. This trend is encouraging more farmers to adopt pasture-raised systems, which offer a viable alternative to industrial pig farming.

ENVIRONMENTAL IMPACT AND SUSTAINABILITY

Pasture-raised pig farming offers a range of environmental benefits that make it a more sustainable alternative to conventional pig farming. This method leverages the natural behaviors of pigs to enhance soil health, support biodiversity, manage waste more effectively, and contribute to carbon sequestration, all of which collectively foster a healthier and more resilient ecosystem.

Soil Health

One of the key advantages of pasture-raised pig farming is the positive impact on soil health. Pigs naturally engage in rooting and foraging behaviors, which involve digging into the soil with their snouts. This activity acts as a natural tilling process, aerating the soil and improving its structure. As pigs forage, they also incorporate organic matter, such as plant residues and their own manure, into the soil. This organic matter decomposes and enriches the soil, increasing its fertility and enhancing its capacity to retain water. Improved soil structure and fertility can lead to more robust plant growth, which in turn supports a healthy ecosystem and sustainable agricultural practices.

Biodiversity

Pasture-based pig farming systems are conducive to supporting a diverse range of plant and animal species. Unlike conventional pig farming, which often involves confinement and monoculture feed crops, pasture systems allow pigs to graze on a variety of plants. This diversified plant cover can provide habitat and food sources for other wildlife, promoting a richer biodiversity. Additionally, rotating pigs through different pasture areas can prevent overgrazing and allow plant species to recover, further enhancing the ecological balance. A diverse ecosystem is more resilient to pests and diseases, reducing the need for chemical interventions and supporting overall ecosystem health.

Waste Management

Effective waste management is another significant benefit of pasture-raised pig farming. In conventional farming systems, large amounts of manure can accumulate in confined spaces, posing a risk of water and soil pollution due to runoff and improper disposal. In contrast, pasture systems distribute pig manure more evenly across the land, where it acts as a natural fertilizer. This reduces the reliance on chemical fertilizers and mitigates the risk of nutrient runoff into waterways. Properly managed manure in pasture systems enhances soil nutrient content and promotes sustainable agricultural practices, contributing to the health of surrounding environments.

Carbon Sequestration

Healthy pastures play a crucial role in carbon sequestration, which is the process of capturing and storing atmospheric carbon dioxide in the soil. Plants in well-managed pastures absorb carbon dioxide through photosynthesis and store it in their roots and the surrounding soil. When pigs graze on these pastures, their activity stimulates plant growth, which can increase the amount of carbon stored in the soil. By enhancing soil carbon content, pasture-raised pig farming can help mitigate climate change by reducing the overall concentration of greenhouse gases in the atmosphere. Additionally, maintaining healthy pastures can prevent soil erosion and degradation, further supporting long-term carbon storage and soil health.

OVERVIEW OF PIG BREEDS SUITABLE FOR PASTURE-RAISING

Pasture-raising pigs offers several advantages, including improved animal welfare, environmental benefits, and high-quality meat production. Choosing the right pig breeds for such systems is crucial, as certain breeds excel in outdoor environments due to their foraging abilities, hardiness, and temperament. Here's a detailed look at some pig breeds that are particularly well-suited for pasture-raised systems:

1. Berkshire

The Berkshire pig is renowned for its marbled, flavorful meat and its ability to thrive in outdoor settings. Originating from England, this breed is hardy and well-adapted to foraging. Berkshires have a calm disposition, which makes them suitable for pasture-based farming where they can roam freely and utilize their natural instincts to find food.

2. Tamworth

Often referred to as "the grazing pig," Tamworths are known for their exceptional foraging abilities. Originating from Ireland and England, these pigs are prized for their efficiency in converting pasture into meat. Tamworths are typically leaner than some other breeds, with a distinctive red coat and a strong build that reflects their robust nature.

3. Large Black

The Large Black pig is characterized by its large size, black coloration, and docile temperament. Originating from England, this breed is well-suited to extensive grazing systems. Large Blacks are known for their excellent foraging skills, which allows them to efficiently utilize pasture resources while maintaining good health and meat quality.

4. Gloucestershire Old Spot

The Gloucestershire Old Spot, or simply Old Spots, is a traditional British breed known for its distinctive white coat with black spots. These pigs are celebrated for their hardiness and adaptability to outdoor conditions. Old Spots have a calm demeanor and are efficient foragers, making them ideal for pasture-raised farming where they can contribute to sustainable meat production.

5. Red Wattle

The Red Wattle pig is recognized for its unique appearance, characterized by a red coloration and distinctive wattles (skin appendages hanging from the neck). Originating from the United

States, this breed is known for its adaptability to various environments, including pasture-based systems. Red Wattles are valued for their high-quality meat and their ability to thrive on a diet that includes pasture vegetation.

☐

CHAPTER TWO

PLANNING AND PREPARATION

ASSESSING LAND AND RESOURCES

Assessing land and resources constitutes the foundational step in establishing a successful farming operation. This critical process involves a comprehensive evaluation of various factors that directly influence agricultural productivity and sustainability.

The primary focus begins with the land itself. Size, quality, and suitability are paramount considerations. The size of the land determines the scale of operations feasible, while quality and suitability assess whether the land can support intended agricultural activities effectively. Key elements such as soil type, topography, drainage patterns, and access to water play crucial roles. Fertile soil is essential for crop cultivation, offering nutrients necessary for plant growth. Proper drainage ensures excess water does not accumulate, preventing waterlogging that can harm crops. Similarly, access to reliable water sources is vital for irrigation, particularly in regions prone to drought or with irregular rainfall patterns.

Moreover, assessing resources extends beyond the land to encompass natural resources that can significantly impact farm operations. Water sources, whether from rivers, lakes, groundwater, or rainwater harvesting systems, are fundamental for sustaining agriculture. Availability and reliability of water sources dictate irrigation strategies and influence crop choices. Renewable energy sources such as solar and wind power offer sustainable alternatives for energy-intensive farming activities, reducing reliance on non-renewable resources and cutting operational costs over time.

Understanding these factors facilitates informed decision-making regarding land use and resource allocation. For instance, land deemed suitable for crop production due to fertile soil and adequate irrigation might be earmarked for intensive farming of cereals or vegetables. Conversely, pastureland with lush grass and sufficient water access may be designated for livestock grazing, optimizing land use according to its inherent capabilities.

Assessment also involves anticipating future needs and challenges. Climate considerations, including rainfall patterns and temperature fluctuations, are critical for long-term planning. Regions experiencing arid conditions might necessitate investments in advanced irrigation systems or drought-resistant crop varieties. Conversely, areas prone to heavy rainfall might require soil conservation measures to prevent erosion and nutrient loss.

Furthermore, regulatory and economic factors influence decision-making.

Compliance with environmental regulations ensures sustainable land management practices, safeguarding natural resources for future generations. Economic viability assessments weigh operational costs against potential yields, guiding investments in infrastructure and technology to enhance productivity and profitability.

CHOOSING THE RIGHT BREED

Choosing the right breed of livestock is a critical decision that significantly influences the success and profitability of a farming operation. Each breed possesses distinct characteristics such as milk production, meat quality, disease resistance, and adaptation to specific climates. Factors that should be carefully weighed include market demand, local climate conditions, available feed resources, and desired production outcomes.

In dairy farming, selecting the appropriate breed can mean the difference between average and exceptional milk yields. Breeds like Holstein and Jersey are highly regarded for their ability to produce large quantities of milk. Holsteins, known for their distinctive black and white markings, are the most common dairy breed globally and excel in high-volume milk production. Jerseys, with their fawn coloration, produce milk with a high butterfat content, making them prized for cheese production and specialty dairy products.

For beef production, considerations lean towards breeds renowned for meat quality and efficient conversion of feed into muscle. Angus and Hereford cattle are particularly favored. Angus, originating from Scotland, are known for their marbled beef, which is tender and flavorful. Herefords, with their distinctive red and white markings, are celebrated for their robustness, adaptability to various climates, and well-marbled beef cuts.

In poultry farming, the choice of breed hinges on whether the focus is on egg production, meat production, or a dual-purpose breed. Rhode Island Reds are esteemed for their consistent egg-laying abilities, producing brown eggs favored by many consumers. On the other hand, for meat production, Cornish Cross chickens are popular due to their rapid growth rate and efficient feed conversion, yielding plump, meaty carcasses in a relatively short time frame.

Beyond productivity metrics, selecting the right breed involves understanding the breed's health considerations and management requirements. Some breeds may be more susceptible to certain diseases or require specific nutritional needs that must be met for optimal health and performance. Climate tolerance is another crucial factor; breeds adapted to local environmental conditions are likely to thrive with fewer health issues and lower management inputs.

Moreover, market dynamics play a pivotal role in breed selection. Understanding consumer preferences and market demand for specific types of meat or dairy products can guide farmers in choosing breeds that align with profitable market niches. Additionally, government regulations and support programs may incentivize the adoption of certain breeds, further influencing the economic viability of a farming enterprise.

The financial implications of breed selection are significant. Initial investment costs for purchasing breeding stock or hatching eggs vary widely depending on the breed's pedigree, age, and genetic characteristics. Ongoing operational costs such as feed, veterinary care, and housing must also be considered in relation to the expected revenue streams from milk, meat, eggs, or other products derived from the chosen breed.

DEVELOPING A BUSINESS PLAN

Developing a comprehensive business plan is essential for any farm, serving as a strategic roadmap that outlines objectives, operational strategies, financial forecasts, and contingency measures. A well-structured plan not only guides day-to-day operations but also plays a crucial role in securing financing and attracting potential investors. Here are the key components typically included in a farm business plan:

Executive Summary: This section provides a concise overview of the farm's mission, goals, and competitive advantages. It encapsulates the essence of the business, highlighting what sets it apart in the market and its long-term vision.

Market Analysis: Conducting a thorough market analysis is crucial for understanding the farm's position within its industry. This involves assessing the target market demographics, demand trends, and the competitive landscape. Insights gained here inform strategic decisions related to product offerings, pricing strategies, and market positioning.

Operational Plan: The operational plan outlines the production processes, resource management strategies, and day-to-day operational workflows. It details how the farm will efficiently produce goods or services, manage inventory, and ensure quality control. This section also addresses regulatory compliance and any necessary certifications.

Financial Plan: The financial plan is a critical component that includes detailed budgets, cash flow projections, and financing requirements. It outlines the expected income and expenses over a specific period, typically covering several years. Financial projections help assess the farm's financial viability and its ability to meet financial obligations and achieve profitability.

Marketing Strategy: A well-defined marketing strategy outlines how the farm will reach its target market and promote its products or services. This includes selecting appropriate sales channels (e.g., direct-to-consumer, wholesalers), setting pricing strategies, and planning promotional activities (e.g., advertising, social media campaigns). The marketing strategy aims to build brand awareness, attract customers, and foster long-term customer relationships.

Risk Management: Identifying potential risks and developing mitigation strategies is crucial for safeguarding the farm's operations and investments. Risks may include natural disasters, market fluctuations, regulatory changes, or operational challenges. The risk management section outlines proactive measures to minimize risks and ensure business continuity.

Each of these components plays a pivotal role in shaping the farm's development and operational framework. The executive summary succinctly encapsulates the farm's mission and competitive advantages, setting the tone for the detailed market analysis that follows. Understanding the target market and competitive landscape informs the operational and marketing strategies, ensuring alignment with market demand and opportunities.

Financial projections underpin strategic decision-making and provide a roadmap for resource allocation and growth. Meanwhile, risk management strategies anticipate potential challenges and prepare the farm to respond effectively, thereby enhancing resilience and sustainability.

LEGAL AND REGULATORY CONSIDERATIONS

In agricultural operations, navigating legal and regulatory considerations is paramount for ensuring compliance, mitigating risks, and sustaining operational integrity. These regulations encompass a spectrum of factors including zoning laws, environmental standards, food safety protocols, and animal welfare guidelines. Understanding and adhering to these requirements are not only legal imperatives but also crucial for maintaining a positive reputation, accessing markets, and promoting sustainable practices.

Zoning laws dictate how land can be used for agricultural purposes, specifying which activities are permitted in designated zones. Compliance ensures that agricultural activities do not conflict with residential, commercial, or industrial areas, thereby preventing legal disputes and preserving community harmony.

Environmental regulations are designed to protect natural resources such as soil, water, and air quality. Farmers must adhere to standards governing pesticide use, waste management, and conservation practices to minimize environmental impact and safeguard ecosystems. Compliance not only avoids fines and penalties but also demonstrates commitment to sustainable agriculture, which is increasingly valued by consumers and stakeholders.

Food safety standards are critical in agricultural production, ensuring that crops and livestock products are safe for consumption. Compliance involves following protocols for handling, processing, and storing food to prevent contamination and outbreaks of foodborne illnesses. Adherence to these standards not only protects public health but also enhances market access by meeting requirements set by retailers, distributors, and regulatory bodies.

Animal welfare guidelines are another key consideration, particularly for livestock operations. These regulations outline standards for housing, handling, and care of animals to ensure their well-being and prevent cruelty. Compliance fosters humane practices and addresses ethical concerns within the farming community and broader society.

Depending on the specific agricultural activities and geographical location, obtaining permits and licenses may be necessary. These can include permits for land use, water rights, pesticide application, and food processing facilities. Meeting these regulatory requirements is essential for operating legally and avoiding disruptions to business operations.

Staying informed about evolving regulations is crucial for agricultural producers. Laws and standards often change in response to scientific advancements, public health concerns, or shifts in societal values. Farmers must proactively monitor these developments and adapt their practices accordingly to maintain compliance and operational efficiency.

Seeking professional advice from legal experts, environmental consultants, or agricultural extension services can provide valuable guidance in navigating complex regulatory landscapes. These professionals offer insights into compliance requirements, assist with permit applications, and help develop strategies for sustainable and legally sound agricultural practices.

BUILDING INFRASTRUCTURE AND FENCING

Building appropriate infrastructure and fencing is crucial for establishing a functional and secure farming environment. These elements not only support the operational needs of farms but also ensure the safety and efficiency of agricultural activities.

Infrastructure encompasses a variety of structures essential for both livestock and crop production. Barns provide shelter for animals and storage for feed and equipment, while sheds and silos offer additional storage capacity for crops and harvested materials. Efficient equipment storage facilities are also integral, safeguarding valuable machinery from the elements and prolonging their operational lifespan.

The design of farm infrastructure is influenced by several key factors. The scale of operations dictates the size and capacity of structures needed to support production requirements effectively. For instance, large-scale farms may require expansive barns and multiple silos, whereas smaller operations might suffice with more modest facilities. Climatic conditions play a crucial role as well; regions prone to extreme weather may necessitate reinforced structures capable of withstanding heavy snow loads or high winds.

Animal welfare standards are paramount in designing infrastructure for livestock. Barns should provide adequate space, ventilation, and bedding to ensure the comfort and health of animals. This not only enhances productivity but also complies with ethical farming practices. Operational efficiency is another critical consideration; strategically located infrastructure reduces the time and effort required to perform daily tasks, thereby optimizing workflow and minimizing labor costs.

Investing in durable materials is essential for the longevity and reliability of farm structures. Steel, concrete, and treated lumber are common choices due to their strength and resilience against weathering. Moreover, incorporating sustainable practices such as energy-efficient lighting and water-saving technologies can yield long-term cost savings and environmental benefits.

Fencing plays a vital role in farm management, serving multiple purposes beyond simple boundary demarcation. It facilitates controlled movement of livestock, preventing them from straying into unwanted areas or neighboring properties. Additionally, well-maintained fencing protects crops from wildlife intrusion, minimizing losses due to grazing or browsing. In the context of biosecurity, fencing helps restrict access to livestock, reducing the risk of disease transmission from wild animals.

Effective fencing design considers factors such as the type of livestock, terrain variability, and the desired level of security. Livestock fencing may range from traditional barbed wire to electric fencing, depending on the species and behavioral characteristics of the animals. In areas prone to wildlife interference, additional measures such as deer fencing or netting may be necessary to safeguard valuable crops.

CHAPTER THREE

PIG HEALTH AND NUTRITION

UNDERSTANDING PIG NUTRITIONAL NEEDS

Pigs, similar to other animals, possess distinctive nutritional requirements crucial for their growth, reproductive health, and overall well-being. Understanding these needs entails familiarity with essential nutrients such as proteins, carbohydrates, fats, vitamins, and minerals. These nutrients play vital roles in muscle development, energy metabolism, immune function, and the maintenance of physiological processes.

The nutritional requirements of pigs can vary significantly depending on several factors including age, breed, weight, and reproductive status. For instance, young, growing pigs have heightened demands for proteins and energy to support their rapid growth rate and development of lean muscle mass. In contrast, lactating sows require diets that are carefully formulated to meet their increased nutritional needs for milk production while maintaining their own body condition.

Proteins are critical components in a pig's diet as they provide the building blocks (amino acids) necessary for muscle growth, tissue repair, and the production of enzymes and hormones. Sources of protein in pig feed often include soybean meal, fish meal, and dried whey, which are carefully balanced to ensure amino acid profiles meet the pig's specific requirements.

Carbohydrates serve as a primary energy source for pigs, supporting daily activities, growth, and reproductive functions. Common sources of carbohydrates in pig diets include grains such as corn, barley, and wheat, which are readily digestible and provide a source of glucose for energy production.

Fats are another essential component of pig diets, supplying concentrated energy and serving as carriers for fat-soluble vitamins (A, D, E, and K). Vegetable oils, animal fats, and by-products like tallow contribute to the fat content of pig feeds, aiding in energy provision and enhancing palatability.

Vitamins and minerals are micronutrients that play crucial roles in various metabolic processes and maintaining overall health. For example, vitamin A supports vision and immune function, while calcium and phosphorus are essential for bone development and structural integrity. Vitamins and minerals are often supplemented in pig diets to ensure optimal growth, reproduction, and disease resistance.

Maintaining optimal nutrition in pigs involves formulating diets that meet their specific requirements at each stage of life. Starter diets for young pigs are designed to promote early growth and digestive health, typically containing higher protein and energy levels. As pigs mature, their nutritional needs change, requiring adjustments in diet composition to support maintenance or reproductive functions.

FORAGE AND SUPPLEMENTAL FEEDING

Forage and supplemental feeding are integral components of optimizing pig nutrition, addressing both their natural dietary needs and ensuring balanced growth and health. Forage, encompassing pasture grasses and legumes, plays a pivotal role by supplying essential fiber and supplementary nutrients that complement the nutritional profile of commercial feeds. This inclusion not only enhances the overall diet's diversity but also supports natural foraging behaviors, contributing to the pigs' physical and mental well-being.

Pasture grasses and legumes are rich sources of various nutrients such as vitamins, minerals, and proteins, which can augment the basic nutritional requirements provided by formulated feeds. This natural supplementation not only enriches the diet but also promotes digestive health through increased fiber intake, which aids in maintaining optimal gut function and overall metabolic efficiency. Moreover, foraging allows pigs to exhibit natural behaviors like rooting and grazing, which are crucial for their physical development and psychological enrichment.

In addition to forage, supplemental feeding plays a crucial role in ensuring pigs receive a balanced diet throughout different growth stages and seasonal variations. Supplemental feeds are formulated to provide specific nutrients that may be deficient in forage or base diets, thereby bridging nutritional gaps and promoting steady growth and development. These feeds are meticulously designed to meet the precise nutritional requirements of pigs at various life stages, from piglets to mature adults, ensuring each phase of growth is adequately supported.

The process of balancing forage and supplements necessitates a nuanced understanding of the nutritional composition of both. Forage quality can vary significantly depending on factors such as plant species, soil health, and seasonal changes, influencing its nutrient content and availability. Therefore, farmers must regularly assess and adjust the forage component of pig diets to maintain consistency and adequacy in nutritional intake.

Similarly, the selection and administration of supplemental feeds require careful consideration of nutritional content, digestibility, and compatibility with existing diet formulations. This ensures that pigs receive a comprehensive array of essential nutrients, including proteins, carbohydrates, fats, vitamins, and minerals, crucial for their overall health and productivity.

Strategically integrating forage and supplemental feeding not only optimizes nutritional intake but also contributes to economic efficiencies by potentially reducing overall feed costs. By leveraging natural resources such as pasturelands effectively, farmers can mitigate dependence on commercial feeds while simultaneously enhancing the sustainability of their operations.

Furthermore, the practice of forage-based and supplemented feeding aligns with broader agricultural goals of sustainability and animal welfare. It supports environmentally friendly farming practices by utilizing natural landscapes for dietary enrichment, reducing the environmental footprint associated with intensive feed production and distribution.

WATER REQUIREMENTS

Water is indispensable for the health and welfare of pigs, playing crucial roles in regulating body temperature, supporting digestion, and facilitating the absorption of nutrients. Access to clean, fresh water is vital for pigs to prevent dehydration and maintain optimal health. Monitoring water intake becomes imperative as requirements can fluctuate depending on environmental conditions, age, and diet composition, ensuring pigs remain hydrated and healthy.

Firstly, water serves as a primary regulator of body temperature in pigs. Through mechanisms such as evaporative cooling, water helps pigs manage heat stress, which is especially critical in hot climates or during periods of intense physical activity. Without adequate water, pigs can quickly succumb to heat exhaustion, leading to severe health complications.

Secondly, water plays a pivotal role in the digestion of food. Pigs require water to produce saliva, which contains enzymes crucial for breaking down food particles. Furthermore, water acts as a solvent, aiding in the digestion and absorption of nutrients from feed. Insufficient water intake can impair digestion, leading to reduced feed efficiency and potentially causing digestive disorders such as constipation or diarrhea.

Moreover, ensuring pigs have constant access to clean water is essential for maintaining their overall health. Clean water prevents the transmission of waterborne diseases and reduces the risk of microbial contamination from pathogens such as bacteria or parasites. Water quality should be regularly monitored to prevent potential health hazards and to ensure pigs are not exposed to harmful substances that could compromise their well-being.

Water requirements vary depending on several factors. Environmental conditions such as temperature and humidity influence how much water pigs need to consume to stay hydrated. For instance, pigs may increase their water intake during hot weather to regulate their body temperature and maintain hydration levels. Similarly, lactating sows or pigs fed high-fiber diets may also have increased water needs compared to pigs under different conditions.

Age is another determinant of water requirements. Young pigs, especially piglets, have higher water needs relative to their body weight compared to adult pigs. As pigs grow, their ability to regulate water balance improves, but adequate hydration remains crucial throughout all stages of life.

Diet composition is yet another factor influencing water intake. Pigs fed diets high in dry matter, such as grains or concentrates, typically require more water to aid in digestion and maintain proper metabolic function. Conversely, pigs on wet or liquid diets may derive a portion of their water intake directly from their feed, affecting their overall drinking water requirements.

To ensure pigs receive sufficient water, it is essential to monitor their water consumption regularly. This can be achieved by observing water trough levels, measuring water flow rates, or using automated systems that record water intake. Any significant deviation from normal water consumption patterns should prompt investigation to identify potential health issues or management concerns.

COMMON HEALTH ISSUES AND PREVENTION

Pigs, like all animals, are prone to a range of health challenges including respiratory diseases, digestive disorders, and issues caused by parasites. Preventing these health problems is paramount in maintaining the well-being and productivity of pig populations. Effective prevention strategies involve implementing rigorous biosecurity measures, ensuring clean and hygienic housing and feeding environments, and closely monitoring the health of pigs on a regular basis.

Biosecurity measures are fundamental in preventing the spread of diseases among pig populations. These measures include controlling access to pig facilities, disinfecting equipment and vehicles entering the premises, and minimizing contact with other animals, wild or domestic, that could potentially carry diseases. By strictly adhering to biosecurity protocols, farmers can significantly reduce the risk of disease transmission within their herds.

Maintaining clean housing and feeding areas is equally crucial. Pigs housed in dirty or overcrowded conditions are more susceptible to stress and illness.

Regular cleaning and disinfection of pens, stalls, and feeding equipment help create a healthier environment for pigs, reducing the likelihood of infections and diseases.

Nutrition plays a pivotal role in supporting the immune system of pigs. Providing a balanced diet that meets their nutritional needs helps in maintaining optimal health and disease resistance. Adequate access to clean water is also essential for proper digestion and overall health.

Early detection of health issues is essential for timely intervention and to prevent minor problems from escalating into major threats. Farmers and caretakers should routinely observe their pigs for signs of illness such as changes in appetite, behavior, or physical appearance. Prompt veterinary care should be sought whenever health concerns arise to diagnose issues early and administer appropriate treatment.

Regular veterinary inspections and health monitoring programs are integral components of proactive pig management. These inspections can identify potential health risks early on and enable preventive measures to be implemented swiftly. Vaccination programs tailored to the specific disease risks of the region or farm can further bolster immunity against prevalent pathogens.

Parasites pose a significant health risk to pigs. Regular deworming and parasite control measures should be implemented as part of a comprehensive health management plan. This helps prevent parasitic infestations that can compromise the health and growth of pigs.

VACCINATION AND VETERINARY CARE

Vaccination plays a pivotal role in maintaining the health and well-being of pigs, safeguarding them against a range of infectious diseases that can pose significant threats to herd stability and productivity. Developing an effective vaccination strategy is crucial and should always be undertaken in collaboration with a qualified veterinarian. This ensures that the approach is tailored to the specific regional disease risks and the unique health history of the herd.

The process of vaccination begins with a comprehensive assessment of the prevalent diseases in the area where the pigs are raised. Different regions may face varying risks from diseases such as swine fever, respiratory infections, or specific viral strains. By understanding these risks, veterinarians can recommend appropriate vaccines and establish a vaccination schedule that maximizes protection without overburdening the animals' immune systems.

Regular veterinary care extends beyond vaccinations to encompass a spectrum of proactive health management practices. Routine health checks are essential to monitor the overall well-being of the pigs. During these examinations, veterinarians assess various indicators of health, including weight, body condition, and signs of illness. Early detection of health issues allows for prompt intervention, which can prevent minor ailments from developing into more serious conditions that could affect the entire herd.

Parasite control is another critical component of veterinary care for pigs. Parasites such as worms can significantly impact the health and growth of pigs, leading to reduced feed efficiency and overall productivity. Veterinarians advise on effective parasite prevention and treatment measures, ensuring that pigs are protected from both internal and external parasites throughout their life cycle.

In addition to disease prevention and parasite control, veterinarians provide valuable management advice to optimize the pigs' living conditions and nutritional needs. Factors such as housing design, ventilation systems, and feeding practices can all influence the health and stress levels of pigs. By tailoring management strategies to the specific requirements of the herd, veterinarians help ensure that pigs thrive in their environment and reach their full potential in terms of growth and reproductive performance.

Furthermore, veterinarians play a crucial role in maintaining biosecurity measures on pig farms. Preventing the introduction and spread of diseases is paramount to the sustainability of pig production. Veterinarians work closely with farm personnel to implement biosecurity protocols, which may include measures such as restricting visitor access, controlling vehicle movement, and maintaining strict hygiene practices.

Collaboration between pig farmers and veterinarians is essential for the success of a comprehensive health management program. Regular communication allows for ongoing assessment and adjustment of strategies based on evolving disease threats, environmental conditions, and herd dynamics. This proactive approach not only protects the health of individual pigs but also contributes to the overall sustainability and profitability of pig farming operations.

□

CHAPTER FOUR

BREEDING AND REPRODUCTION

SELECTING BREEDING STOCK

Selecting the right breeding stock is a critical decision in pig farming, directly influencing the future quality and profitability of the herd. Breeding stock refers to pigs chosen specifically for their ability to pass on desirable traits to offspring, ensuring improvements in health, size, and productivity with each generation.

The process of selecting breeding stock involves careful consideration of several key factors. Firstly, health and genetics are paramount. Healthy pigs free from hereditary diseases form the foundation of a robust breeding program. Beyond health, selecting pigs with superior genetics is crucial. These genetics contribute to traits such as rapid growth, high-quality meat production, and efficient reproduction, all of which are essential for a successful pig farming operation.

Conformation, or the physical structure of the pig, is another significant consideration. Well-built pigs with balanced proportions are more likely to produce offspring that thrive. They are generally healthier, grow faster, and utilize feed more efficiently, which contributes to overall farm productivity.

Temperament plays a role as well. Calm and manageable pigs are easier to handle during breeding, farrowing (giving birth), and general husbandry practices. A gentle temperament reduces stress on both the animals and the farmers, enhancing overall farm efficiency and welfare.

Productivity is a key factor in selecting breeding stock. Pigs that consistently produce large litters and successfully wean piglets are highly desirable. This trait directly impacts the farm's profitability by increasing the number of market-ready pigs per breeding cycle and maximizing overall output.

Farmers utilize various tools and methods to assess potential breeding stock effectively. Performance records are invaluable, providing quantitative data on traits such as litter size, growth rate, feed efficiency, and maternal instincts. These records allow farmers to identify pigs that consistently perform well across these metrics, guiding decisions to retain or cull specific animals from the breeding program.

Genetic evaluations complement performance records by providing insights into the heritability of traits. This analysis helps farmers predict the potential genetic contribution of individual pigs to future generations, aiding in the selection of breeding stock that aligns with long-term breeding goals.

Furthermore, pedigree analysis can trace ancestry and identify genetic lines known for desirable traits. This historical data informs breeding decisions, allowing farmers to capitalize on established genetic strengths within their herd.

UNDERSTANDING THE BREEDING CYCLE

Understanding the breeding cycle of pigs, particularly for successful mating and reproduction, is crucial for farmers and breeders. Female pigs, known as sows, follow a regular estrous cycle that typically spans about 21 days. This cycle is divided into key stages that play significant roles in the reproductive process.

The first critical stage is Estrus (Heat). Estrus is the period when the sow becomes sexually receptive and shows signs that she is ready to mate. This phase typically lasts between 1 to 3 days. Signs of estrus include restlessness, increased vocalization, mounting behavior (either on other pigs or objects), and a swollen vulva. During this time, the sow releases pheromones that signal her readiness to potential mates, making this period crucial for timing mating attempts.

Following estrus is Ovulation, which occurs approximately 40 to 45 hours after the onset of estrus. Ovulation is the release of mature eggs from the sow's ovaries into the fallopian tubes, where they await fertilization by sperm. Timing mating to coincide with ovulation is critical because it maximizes the chances of successful fertilization. Sperm can survive in the female reproductive tract for several days, so mating slightly before ovulation can still result in fertilization.

If fertilization is successful, the next stage is Gestation. Gestation is the period during which the sow carries and develops the embryos/fetuses until they are ready to be born as piglets. The gestation period for pigs lasts approximately 3 months, 3 weeks, and 3 days, which is about 114 days in total. During gestation, proper nutrition and care are essential to ensure the health of both the sow and the developing piglets.

Farmers employ various methods to detect estrus in sows to optimize the timing of mating. Visual Observation is one of the oldest and simplest methods, where farmers closely observe the behavior and physical signs of the sow. This includes noting changes in activity level, vocalizations, and the appearance of the vulva. Another method involves Boar Exposure, where sows are placed near a boar. The presence of a boar can stimulate sows to display signs of estrus, such as standing still when pressure is applied to their backs (a behavior known as the "standing reflex"). This method takes advantage of natural behaviors triggered by pheromones emitted by the boar.

In recent years, Electronic Heat Detection Systems have become increasingly popular on farms. These systems utilize technology such as pedometers, activity monitors, or sensors that detect changes in behavior or physiological markers associated with estrus. They provide a more objective and continuous monitoring of estrus signs, which can enhance the accuracy of timing for insemination.

GESTATION AND FARROWING

Gestation and farrowing are critical stages in the lifecycle of pigs, demanding careful attention and management to ensure the health and well-being of both sows and piglets.

Gestation:

Gestation in pigs refers to the period of pregnancy, lasting approximately 114 days (3 months, 3 weeks, and 3 days). During this phase, proper nutrition is crucial for the sow to support the developing litter. Sows in gestation require a balanced diet that meets their nutritional needs while also supporting the growth of the piglets. Typically, they are fed a combination of grains, protein supplements, vitamins, and minerals to ensure optimal health and growth. It's essential to monitor their body condition and adjust their diet as needed to prevent issues such as obesity or malnutrition, which can affect the sow's health and the development of the piglets.

Aside from nutrition, sows in gestation require a comfortable and stress-free environment. They are often housed in gestation crates or pens that allow for individual care and monitoring. These facilities should be well-maintained and provide ample space for the sow to move comfortably.

Regular health checks during gestation are essential to identify any potential issues early. This includes monitoring for signs of illness or discomfort, as well as ensuring vaccinations and parasite control measures are up to date. Proper management practices help minimize stress and maximize the sow's reproductive efficiency.

Farrowing:

Farrowing is the natural process of giving birth to piglets. It typically occurs in a specially prepared farrowing crate or pen. These environments are designed to provide a safe and hygienic space for the sow and her offspring. Farrowing crates have raised edges or rails to prevent the sow from accidentally crushing her piglets, a common concern during the first few days post-birth.

While sows usually manage farrowing without assistance, farmers should be prepared to intervene if necessary. This may involve helping with the delivery of piglets or providing emergency care to the sow or newborns. Experienced farmers often monitor sows closely as they approach their due date, ensuring everything is in place for a smooth farrowing process.

After farrowing, the immediate focus shifts to ensuring the health and survival of the newborn piglets. Colostrum, the sow's first milk, is crucial as it provides antibodies and essential nutrients that boost the piglets' immune systems and help them thrive. Piglets must nurse within the first few hours of birth to receive this vital colostrum. Farmers may assist by ensuring each piglet has access to a teat and that weaker piglets receive additional care if needed.

Maintaining a warm environment is also essential during the fragile first days of life. Piglets lack the ability to regulate their body temperature effectively, so supplemental heating or heat lamps may be used to keep them comfortable. Additionally, regular health checks for the piglets help identify and address any health issues early on.

CARE OF PIGLETS

Proper care of piglets is essential to ensure their survival and healthy development. From birth onwards, piglets require careful attention and specific interventions to support their growth.

Immediately after birth, piglets are particularly vulnerable and reliant on colostrum, the first milk produced by the sow. Colostrum is crucial as it provides essential nutrients and antibodies that help piglets develop immunity to diseases. It's vital for farmers to ensure each piglet receives colostrum within the first few hours after birth, as the ability to absorb antibodies diminishes rapidly after the first 24 hours.

In some cases, weaker piglets may struggle to nurse adequately. Farmers may need to intervene by providing supplemental milk to ensure these piglets receive the necessary nutrition. This intervention is critical to preventing malnutrition and giving weaker piglets a better chance at survival.

Temperature regulation is another critical aspect of piglet care. Newborn piglets have limited ability to regulate their body temperature and are prone to chilling, which can be fatal if not addressed promptly. Maintaining a warm environment is essential, especially during the first few days of life. Heat lamps or heated mats are commonly used to provide a warm area where piglets can rest and stay comfortable.

As piglets grow, their nutritional needs change. They begin to transition from solely relying on milk to consuming solid feed. This transition typically starts with creep feed, which is specially formulated to be easily digestible for young piglets. Creep feed allows piglets to gradually adapt to solid food and provides additional nutrients necessary for their development.

Weaning marks a significant milestone in a piglet's life. It generally occurs around 3-4 weeks of age, although the timing can vary depending on farm practices and the health status of the piglets. Weaning involves separating piglets from the sow and transitioning them entirely to solid feed. This process is carefully managed to minimize stress and ensure the piglets continue to grow and thrive post-weaning.

Throughout the entire process, regular health checks and monitoring are essential. Farmers observe piglets for signs of illness or distress, ensuring prompt intervention if any issues arise.

Preventive measures such as vaccinations are also crucial to safeguarding the health of the piglets and preventing the spread of diseases within the herd.

GENETIC MANAGEMENT AND IMPROVEMENT

Genetic management plays a crucial role in modern livestock farming, particularly in enhancing desirable traits within herds over successive generations. This strategic approach involves meticulous breeding decisions aimed at improving various aspects such as disease resistance, growth rate, meat quality, and overall productivity. By leveraging advanced techniques like genetic markers, artificial insemination

(AI), and embryo transfer, farmers can accelerate genetic progress significantly, often using superior genetics sourced from a select few elite animals.

At the core of genetic management is the careful selection of breeding stock based on their genetic profiles. Genetic markers, identifiable sequences within the DNA associated with specific traits, serve as powerful tools in this process. Farmers can identify animals with desirable traits such as disease resistance or efficient growth, enabling them to make informed decisions about which individuals to breed. This selective breeding approach not only enhances the quality of the offspring but also ensures a steady improvement in herd characteristics over time.

Technological advancements like AI and embryo transfer further expedite genetic improvement. AI allows farmers to use semen from high-performing males, even if those males are not physically present on the farm. This widens the genetic pool accessible to farmers and increases the likelihood of breeding animals with superior traits. Similarly, embryo transfer enables the propagation of genetic material from elite females by transferring embryos into surrogate mothers. These techniques collectively enable farmers to maximize the genetic potential of their herds efficiently.

A critical aspect of genetic management is the systematic collection and analysis of performance data. By regularly evaluating traits such as growth rates, feed efficiency, and health parameters, farmers can track the genetic progress within their herds. This data-driven approach helps in identifying top-performing individuals that should be prioritized for breeding. Over time, continuous monitoring and adjustment of breeding strategies based on performance outcomes are essential to achieving sustained genetic improvement.

The long-term nature of genetic improvement requires farmers to adopt a strategic approach. Careful planning, informed decision-making, and ongoing monitoring are necessary to achieve desired outcomes in productivity and profitability. Farmers must consider multiple factors, including market demands, environmental conditions, and consumer preferences, when selecting breeding goals. This holistic approach ensures that genetic improvements align with broader farm management objectives, ultimately enhancing overall efficiency and sustainability.

CHAPTER FIVE

DAILY CARE AND MANAGEMENT

FEEDING STRATEGIES

Feeding strategies play a pivotal role in livestock management, serving as a cornerstone in meeting animals' nutritional requirements effectively. This section delves into the strategic planning and implementation of feeding routines, meticulously designed to cater to the distinct dietary needs of various livestock species. Key considerations encompass the nutritional composition of feeds, optimal feeding schedules, and efficient methods of feed delivery. Achieving a balanced diet is critical for fostering growth, ensuring reproductive health, and bolstering overall well-being, underscoring the necessity of comprehending dietary demands and fine-tuning feeding practices accordingly.

Central to any feeding strategy is the careful selection of feed types based on their nutritional profile. Livestock species vary widely in their dietary needs, influenced by factors such as age, physiological state (e.g., lactation, pregnancy), and specific production goals (e.g., meat production, milk yield). For instance, dairy cows require diets rich in energy and protein to sustain milk production, while growing calves need diets that support skeletal development and muscle growth.

Equally important is the establishment of feeding schedules that align with animals' metabolic rhythms and production cycles. Consistency and timing in feeding help maintain digestive health and optimize nutrient utilization. For example, ruminants like cattle benefit from multiple feedings throughout the day to support their unique digestive processes, which involve fermentation in the rumen.

The method of feed delivery also impacts feeding efficiency and animal well-being. Technologies such as automated feeders and precision feeding systems have revolutionized modern livestock management, allowing for precise control over feed distribution.

This not only minimizes wastage but also ensures that each animal receives its required nutritional intake. In extensive systems, where animals graze on pasture, rotational grazing strategies can be employed to manage grazing pressure and optimize nutrient availability in the pasture.

Moreover, adapting feeding practices to seasonal variations and environmental conditions is crucial. In regions with distinct seasons, the availability and quality of forage and supplemental feeds fluctuate, necessitating adjustments in feeding strategies to maintain optimal nutrition year-round. Additionally, considerations such as water availability and quality play a pivotal role in ensuring animals can effectively utilize the nutrients they consume.

Beyond nutritional adequacy, feeding strategies contribute significantly to overall herd health and productivity. Monitoring feed intake and body condition scores allows for early detection of nutritional deficiencies or health issues, enabling timely interventions. Furthermore, integrating sustainable practices into feeding strategies, such as sourcing locally produced feeds or reducing reliance on imported supplements, aligns with broader environmental and economic goals.

ROTATIONAL GRAZING PRACTICES

Rotational grazing is a strategic method used in pasture management where grazing land is divided into smaller sections. Livestock are moved through these sections in a planned sequence, ensuring they have access to fresh pasture while allowing previously grazed areas time to recover. This practice offers several benefits crucial for sustainable agriculture, including preventing overgrazing, promoting even forage utilization, and minimizing soil erosion.

The core principle of rotational grazing lies in its ability to mimic natural grazing patterns of wild herbivores. By dividing the pasture into smaller paddocks or sections, farmers can control where and for how long livestock graze. This control is essential for maintaining pasture health and productivity over the long term. When animals are rotated through paddocks, they graze selectively, consuming a portion of the available forage while leaving enough vegetation behind for rapid recovery.

Implementing a successful rotational grazing system involves careful planning and management. Farmers must consider factors such as stocking densities, grazing periods, and rest periods for each paddock. Stocking densities should be adjusted based on the carrying capacity of the land and the specific nutritional needs of the livestock. Grazing periods are typically short to prevent overgrazing, while rest periods allow vegetation time to regrow and replenish root reserves, which are crucial for overall pasture health.

One of the primary benefits of rotational grazing is its positive impact on soil health. By avoiding continuous grazing pressure, the soil structure remains intact, preventing compaction and promoting better water infiltration and nutrient cycling. This healthier soil supports diverse microbial communities that contribute to nutrient availability and overall ecosystem resilience.

Additionally, rotational grazing can enhance forage productivity. When pastures are managed with rotations, plants have the opportunity to recover fully between grazing events. This leads to increased biomass production and higher-quality forage, which is beneficial for livestock nutrition and performance. Farmers often observe improved weight gains and reproductive performance in livestock managed under rotational grazing compared to continuous grazing systems.

From a sustainability perspective, rotational grazing supports efficient land use. By maximizing the productivity of existing pastureland, farmers can reduce the need to expand into natural habitats, thereby conserving biodiversity and ecosystem services. The practice also contributes to greenhouse gas mitigation through improved carbon sequestration in healthy soils and reduced methane emissions associated with improved forage digestion in livestock.

Successful implementation of rotational grazing requires ongoing monitoring and adaptation to local conditions and seasonal changes. Farmers may adjust grazing schedules based on weather patterns, pasture growth rates, and animal behavior to optimize outcomes. Education and support from agricultural extension services and peer networks can also play a crucial role in helping farmers adopt and refine rotational grazing practices.

WASTE MANAGEMENT

Proper waste management plays a pivotal role in safeguarding environmental integrity and curbing disease transmission within livestock farms. This section delves into various strategies aimed at managing manure, wastewater, and other agricultural wastes effectively. By focusing on composting organic waste to yield nutrient-rich fertilizer and implementing runoff control measures to preserve water quality, these practices uphold environmental stewardship while bolstering the sustainability of livestock operations.

Manure, a primary agricultural waste, requires meticulous handling to prevent environmental degradation and maintain farm hygiene. Livestock operations often generate substantial quantities of manure rich in nutrients like nitrogen and phosphorus. Improper disposal can lead to nutrient runoff into water bodies, exacerbating eutrophication and harming aquatic ecosystems. To mitigate these risks, farms employ strategies such as composting. Composting not only decomposes organic waste into stable, humus-like material but also reduces pathogens, thereby producing a valuable soil amendment that enhances fertility and structure.

Composting involves carefully managing organic materials under controlled conditions to promote microbial decomposition. This process requires adequate aeration, moisture, and carbon-to-nitrogen ratios to ensure efficient breakdown and pathogen reduction. By converting waste into beneficial compost, farms minimize waste volumes while recycling nutrients back into the soil, fostering sustainable agricultural practices.

In addition to managing solid wastes like manure, farms must address wastewater generated from various activities such as cleaning pens and processing facilities. Wastewater often contains organic matter, nutrients, and potential contaminants that, if not managed properly, can pollute water sources. Farms utilize systems like sedimentation ponds, vegetated buffer strips, and constructed wetlands to treat wastewater before discharge. These natural filtration methods trap sediments, absorb nutrients, and promote biological degradation, thereby safeguarding water quality and complying with environmental regulations.

Runoff control is another critical aspect of waste management on farms. Rainwater and irrigation runoff can carry pollutants such as nutrients, pesticides, and sediments into nearby water bodies, posing ecological risks. Farms employ practices like contour plowing, grassed waterways, and riparian buffers to minimize runoff and soil erosion. These measures help retain water on fields, reduce sediment transport, and enhance soil stability, preserving both land productivity and aquatic habitats.

By adopting comprehensive waste management practices, livestock farms not only mitigate environmental impacts but also optimize resource efficiency and operational sustainability. Effective waste management reduces greenhouse gas emissions associated with decomposition processes, mitigates odor issues, and minimizes the spread of diseases. Moreover, by converting waste into valuable resources like compost and reducing reliance on synthetic fertilizers, farms contribute to a circular economy that promotes long-term environmental health and agricultural resilience.

WEATHER AND SHELTER CONSIDERATIONS

Providing suitable shelter is crucial for safeguarding livestock from the rigors of weather extremes, ensuring their well-being and productivity remain optimal. This section delves into essential considerations for designing and managing shelters that effectively shield animals from heat, cold, wind, and precipitation. By addressing factors such as ventilation, insulation, bedding materials, and spatial requirements tailored to the size and type of livestock, farmers can mitigate stress-related health problems and promote year-round comfort and productivity.

Livestock, ranging from cattle to poultry, are highly vulnerable to adverse weather conditions. To counteract these challenges, shelter design plays a pivotal role. Proper ventilation is essential to maintain air quality and regulate temperature within the shelter. It prevents the buildup of harmful gases and excessive humidity, which can compromise respiratory health and overall comfort. Strategic placement of openings and vents facilitates natural airflow while protecting animals from drafts, ensuring a conducive environment across seasons.

Insulation serves as a critical barrier against temperature extremes. Depending on the climate, shelters may require insulation to retain warmth during cold spells or reflect heat during scorching temperatures. Materials such as straw, hay, or specialized insulating foams can be employed to bolster thermal efficiency, reducing energy expenditure and stress on livestock. Effective insulation also aids in controlling moisture levels, preventing dampness that could lead to health issues like frostbite or fungal infections.

The choice of bedding material is another pivotal consideration. It should provide comfort, absorb moisture, and maintain cleanliness. Common options include straw, sawdust, or rubber mats, tailored to the specific needs and behaviors of the livestock. Clean and dry bedding reduces the risk of infections, enhances resting conditions, and supports overall health and productivity. Space within shelters must accommodate the size and type of livestock to ensure freedom of movement and social interactions. Overcrowding can exacerbate stress and increase the likelihood of injuries or diseases spreading. Adequate space allocation promotes natural behaviors such as lying down, standing, and grooming, contributing to the animals' physical and mental well-being.

Regular maintenance of shelters is paramount to sustaining their effectiveness. This includes inspecting for structural integrity, repairing damage promptly, and cleaning bedding areas regularly. Proper upkeep extends the longevity of the shelter and preserves its functionality in safeguarding livestock against environmental stressors.

RECORD KEEPING AND MONITORING

Record keeping and monitoring play pivotal roles in the management of livestock, ensuring their health, productivity, and overall performance can be effectively tracked and optimized over time. These practices involve maintaining precise records of various aspects such as feeding schedules, health treatments, breeding activities, and production metrics. Such meticulous documentation serves as a crucial tool in making informed decisions and identifying trends or issues that demand attention within the livestock operation.

Accurate record keeping is essential for several reasons. Firstly, it provides a comprehensive history of each animal, detailing its health status, vaccinations, and any medical treatments administered. This historical data is invaluable in maintaining health records and in complying with regulatory requirements. Additionally, records of feeding schedules and nutrition intake help ensure that each animal receives appropriate nourishment tailored to its specific needs, thereby promoting optimal growth and development.

Moreover, tracking breeding activities through detailed records enables livestock managers to monitor reproductive performance and plan breeding cycles effectively. This is vital for maintaining genetic diversity and improving the quality of livestock over successive generations. Production metrics such as milk yield, egg production, or weight gain are also meticulously recorded to assess overall productivity and profitability of the operation.

Methods for collecting and organizing these records vary, ranging from traditional manual logbooks to modern digital tools. Manual recording may involve daily entries by hand, documenting observations and treatments directly in designated notebooks or forms. While this method can be straightforward, it may be more prone to errors or loss if not managed meticulously. On the other hand, digital tools such as specialized software or apps offer advantages in terms of data accuracy, accessibility, and the ability to generate reports or analyze trends efficiently. Many farmers now opt for digital solutions that integrate data from various aspects of livestock management into a centralized system, enhancing overall operational efficiency.

Regular monitoring of these records allows for early detection of health issues or inefficiencies in productivity. By analyzing trends over time, managers can identify patterns that may indicate emerging health problems or suboptimal management practices. Timely intervention based on these insights can prevent potential losses and optimize the overall health and well-being of the livestock.

In essence, effective record keeping and monitoring are not merely administrative tasks but critical components of successful livestock management. They provide a foundation for informed decision-making, ensuring that resources are allocated efficiently and that livestock are cared for in a manner that maximizes their potential. By maintaining accurate records and leveraging them through regular monitoring, farmers can uphold high standards of animal welfare, productivity, and profitability within their operations. Thus, embracing robust record-keeping practices and leveraging modern monitoring techniques are indispensable steps towards achieving sustainable and efficient livestock management practices in today's agricultural landscape.

CHAPTER SIX

PROCESSING AND MARKETING

HUMANE SLAUGHTER PRACTICES

Humane slaughter practices are paramount in upholding ethical standards and ensuring that animals are treated with dignity and compassion throughout the process of preparing them for meat production. The goal is to minimize stress and pain, thereby mitigating any unnecessary suffering for the animals involved.

Central to humane slaughter practices is the technique of stunning. Stunning is a method used to render animals unconscious before slaughter, thereby preventing them from experiencing pain during the actual killing process. There are various methods of stunning employed, depending on the species and regulations in place. These methods include electrical stunning, captive bolt stunning, gas stunning, and stunning with percussive blow instruments. Each method aims to induce unconsciousness swiftly and effectively, ensuring that the animal is not aware of its surroundings or the impending process.

Regulations and guidelines play a crucial role in overseeing and enforcing humane slaughter practices. They are designed to uphold ethical standards and legal requirements, ensuring that all stages of slaughter—from arrival at the facility to the final moments—are conducted in a manner that respects animal welfare. These regulations often dictate specific procedures for stunning, handling, and slaughter to minimize distress and uphold humane treatment.

Before the slaughter process begins, animals should be handled calmly and gently to reduce stress. Facilities are designed with animal welfare in mind, providing adequate space, ventilation, and non-slip flooring to ensure a calm environment. Proper handling techniques are crucial in minimizing fear and anxiety among the animals.

Once animals are prepared for slaughter, the stunning process begins. The choice of stunning method depends on factors such as the size of the animal, the species, and facility capabilities.

Electrical stunning involves applying a current through the brain or heart, instantly inducing unconsciousness. Captive bolt stunning uses a device to deliver a precise blow to the head, causing immediate unconsciousness through blunt force trauma. Gas stunning involves exposing animals to gases like carbon dioxide or argon, which induce unconsciousness through anoxic conditions.

Following stunning, the animals are swiftly moved to the slaughter stage, where bleeding out occurs. This process is vital for both food safety and animal welfare, ensuring that death is quick and irreversible. Proper bleeding out ensures that the animal loses consciousness due to lack of blood flow to the brain, following humane guidelines to minimize suffering.

Humane slaughter practices are continuously evolving with advancements in technology and understanding of animal behavior and welfare. Efforts are ongoing to improve methods of stunning, handling, and overall slaughter processes to further minimize stress and enhance animal welfare standards. By adhering to these practices and regulations, stakeholders in the meat production industry uphold their commitment to ethical treatment of animals, ensuring that the journey from farm to table is as respectful and compassionate as possible.

MEAT PROCESSING AND PACKAGING

Meat processing is a multifaceted operation integral to converting livestock into consumable products ready for market distribution. This intricate process encompasses several key stages, each crucial in ensuring the meat is not only safe and of high quality but also tailored to meet consumer preferences.

The journey of meat processing begins with the careful selection and inspection of livestock. This initial step is critical for ensuring that only healthy animals are processed, thereby safeguarding the quality and safety of the final meat products. Once selected, the animals undergo slaughter in controlled environments adhering to strict regulatory standards. This humane process is designed to minimize stress and ensure the humane treatment of the animals.

Following slaughter, the carcasses are subjected to rigorous sanitation procedures to eliminate any potential contaminants. This cleanliness is paramount in preventing the spread of pathogens and ensuring the meat's safety. After sanitation, the meat undergoes a series of cutting and trimming processes. These processes not only separate the meat into different cuts but also remove excess fat and connective tissue to enhance both the aesthetic appeal and the eating quality of the meat.

Curing is another significant stage in meat processing, particularly for products like bacon, ham, and sausages. Curing involves the addition of salts, sugars, and sometimes spices to the meat to enhance flavor and prolong shelf life. This traditional preservation method has evolved over centuries and remains a staple in meat processing, offering consumers a range of flavorful options.

Packaging plays a pivotal role in the meat processing industry, serving multiple functions beyond mere containment. Modern packaging methods such as vacuum-sealing and modified atmosphere packaging (MAP) are employed to extend shelf life by minimizing exposure to oxygen, which can accelerate spoilage. These methods also help maintain the meat's freshness and nutritional value while protecting it from contamination during storage and transport.

Moreover, packaging serves as a means of communicating essential information to consumers. Labels provide details such as nutritional content, ingredient lists, and cooking instructions, empowering consumers to make informed choices. Additionally, packaging contributes to the overall presentation of the product, influencing consumer perception and purchasing decisions.

Throughout these processes, stringent quality control measures are implemented to ensure that all products meet regulatory standards and consumer expectations. Regular inspections and testing protocols are conducted to monitor factors such as temperature control, hygiene practices, and microbial levels, thereby mitigating potential risks to food safety.

DIRECT MARKETING STRATEGIES

Direct marketing strategies are integral to selling products directly to consumers, bypassing intermediaries such as retailers. This approach opens up avenues like farmers' markets, community-supported agriculture (CSA) programs, and online platforms, facilitating direct engagement between producers and consumers. The benefits are manifold, ranging from building robust consumer relationships to gaining immediate feedback and potentially achieving higher profit margins compared to traditional wholesale channels.

One of the primary advantages of direct marketing is the ability for producers to establish direct connections with their customers. By participating in farmers' markets or running CSA programs, producers can interact face-to-face with consumers, fostering trust and loyalty.

This personal interaction allows for direct feedback on products, which can be invaluable for refining offerings and meeting consumer preferences effectively.

Moreover, direct marketing enables producers to retain greater control over pricing and margins. By eliminating middlemen, they can capture a larger share of the revenue that would otherwise be diluted through wholesale distribution. This financial benefit is particularly significant for small-scale producers who may struggle to compete in larger retail environments dominated by established brands.

Online platforms further expand the reach of direct marketing strategies, providing producers with a global marketplace accessible 24/7. Websites, e-commerce platforms, and social media channels allow producers to showcase their products directly to consumers worldwide, breaking down geographical barriers and reaching niche markets that might be underserved by traditional retail outlets.

Successful direct marketing strategies hinge on effective branding and customer engagement. Establishing a strong brand identity helps differentiate products in a competitive market landscape. Whether through distinctive packaging, consistent messaging, or a compelling story behind the product, branding plays a pivotal role in attracting and retaining customers.

Furthermore, customer engagement is key to sustaining long-term relationships. Producers can engage with consumers through educational content, newsletters, or loyalty programs, keeping them informed about product updates, seasonal offerings, or upcoming events. This continuous engagement not only strengthens brand loyalty but also encourages repeat purchases and word-of-mouth referrals within the community.

Understanding local market preferences is another critical aspect of effective direct marketing. Producers can tailor their offerings to cater to specific tastes, dietary preferences, or cultural nuances prevalent in their target market. This localized approach enhances relevance and appeal, resonating more deeply with consumers and increasing the likelihood of sales success.

SELLING TO RETAIL AND WHOLESALE MARKETS

Selling products in both retail and wholesale markets entails navigating distinct channels that cater to different consumer needs and distribution dynamics. Retail markets directly serve consumers through grocery stores, specialty shops, and restaurants, necessitating adherence to specific packaging and labeling standards tailored to end-user preferences and regulatory requirements. On the other hand, wholesale markets involve larger-scale transactions where products are sold in bulk to intermediaries who distribute them further to retail outlets.

In retail markets, the consumer-facing nature demands meticulous attention to product presentation, packaging, and labeling. These aspects not only serve practical purposes such as information dissemination and compliance with health and safety regulations but also play a crucial role in attracting and informing customers. For instance, clear and attractive packaging can enhance shelf appeal and facilitate consumer decision-making, while accurate labeling ensures transparency regarding ingredients, nutritional content, and any relevant allergen information. Moreover, packaging must often meet environmental sustainability standards, reflecting growing consumer concerns and regulatory trends towards eco-friendly practices.

Establishing and nurturing relationships with retail buyers is pivotal for success in this market. Effective communication and responsiveness to buyer preferences and market trends are essential. Understanding consumer behavior, regional preferences, and seasonal demands allows suppliers to adapt their offerings accordingly, ensuring products resonate with target demographics and remain competitive in a crowded marketplace.

Conversely, wholesale markets involve transactions on a larger scale, typically with distributors or wholesalers who purchase goods in bulk from manufacturers or suppliers. These intermediaries then distribute the products to various retail establishments. Key considerations in wholesale transactions include negotiating pricing, volume discounts, and logistics arrangements to optimize efficiency and cost-effectiveness throughout the supply chain. Suppliers must also manage inventory levels and production schedules to meet the demands of wholesalers while maintaining consistent product quality and availability.

Building and maintaining strong relationships with wholesalers is critical in wholesale markets. Suppliers often collaborate closely with distributors to forecast demand, plan inventory levels, and coordinate promotional efforts. This partnership-oriented approach helps align supply with market demand, minimize stockouts, and capitalize on sales opportunities.

Across both retail and wholesale markets, ensuring consistent product quality is non-negotiable. Quality control measures, including rigorous testing, adherence to industry standards, and continuous improvement initiatives, are essential to uphold brand reputation and customer trust. In retail settings, where direct consumer interaction occurs, product quality directly impacts customer satisfaction and loyalty. Meanwhile, in wholesale contexts, reliability in product quality fosters trust and long-term partnerships with distributors, enhancing supply chain efficiency and profitability.

VALUE-ADDED PRODUCTS

Value-added products in agriculture represent a strategic transformation of raw commodities into goods that offer enhanced convenience, nutritional value, or appeal to consumers. This process not only distinguishes a producer's offerings in the market but also positions them to command premium prices and explore new market avenues. However, pursuing value-added products necessitates specialized processing skills, investments in equipment, and a keen understanding of consumer preferences.

At its core, the concept of value-added products involves adding value through processing. This can range from transforming basic agricultural commodities like meats, fruits, or grains into ready-to-eat meals, pre-marinated meats, or specialty cuts that cater to modern consumer demands for convenience and quality. For instance, pre-packaged salads, artisanal cheeses, or organic jams are all examples of products that undergo value addition to meet specific consumer preferences and market niches.

The benefits of producing value-added products are manifold. Firstly, it allows agricultural producers to differentiate themselves in a competitive market by offering unique products that are not easily replicated. This differentiation often translates into higher profit margins as consumers are willing to pay more for products that offer added convenience or quality. Moreover, entering the value-added market can mitigate price fluctuations associated with raw commodities by creating products with a stable and potentially higher market value.

However, embarking on value-added production requires careful consideration and investment. Producers must possess or develop the necessary processing expertise to ensure product safety, quality, and consistency. This may involve acquiring new equipment, implementing stringent quality control measures, and complying with regulatory standards governing food processing and safety.

Understanding consumer preferences is also crucial. Market research becomes essential to identify trends, preferences for specific flavors or ingredients, packaging formats, and nutritional considerations. For instance, consumers increasingly seek out products that are organic, locally sourced, or cater to dietary restrictions such as gluten-free or vegan diets. Tailoring value-added products to meet these preferences can significantly enhance market acceptance and profitability.

Furthermore, effective marketing and distribution strategies are pivotal in successfully launching value-added products. Producers may need to forge partnerships with retailers, participate in farmers' markets or specialty food fairs, and leverage digital platforms for direct-to-consumer sales. Building brand recognition and consumer trust through transparent labeling, certifications, and storytelling around the product's origins and benefits can also enhance market penetration and consumer loyalty.

CHAPTER SEVEN

FINANCIAL MANAGEMENT AND ECONOMICS

COST ANALYSIS AND BUDGETING

Cost analysis and budgeting are foundational practices in financial management, essential for businesses and organizations to maintain financial health and achieve strategic objectives. These processes provide crucial insights into the financial operations and resource allocation of an entity, facilitating informed decision-making and sustainable growth.

Cost Analysis: Cost analysis involves a comprehensive examination of the expenses incurred by an organization in its operations. These expenses encompass both direct costs, such as raw materials and labor directly involved in production, and indirect costs like administrative overhead, utilities, and other support functions necessary for business operations. Through meticulous analysis, businesses gain a detailed understanding of their cost structure, identifying areas where costs can be reduced, efficiencies can be improved, or investments can be strategically made.

By scrutinizing costs, organizations can pinpoint inefficiencies or areas of overspending. For instance, identifying high-cost suppliers or inefficient production processes allows businesses to renegotiate contracts or streamline operations, thereby improving profitability. Moreover, cost analysis enables businesses to make informed decisions about pricing strategies, product profitability, and resource allocation, ensuring financial sustainability in competitive markets.

Budgeting: Budgeting is a systematic process of planning and allocating financial resources based on anticipated revenues and expenses over a defined period, typically a fiscal year. It serves as a financial roadmap, guiding organizations in managing cash flow, controlling expenditures, and achieving financial objectives. A well-developed budget aligns financial resources with strategic priorities, ensuring that funds are allocated to activities that support growth and profitability.

The budgeting process begins with setting clear financial goals and objectives, such as revenue targets, cost reduction initiatives, or investment priorities. It involves forecasting revenues from sales, investments, and other sources, as well as estimating various categories of expenses, including operational costs, capital expenditures, and debt servicing. By comparing actual performance against budgeted figures, organizations can assess their financial health, identify variances, and take corrective actions as needed.

Effective budgeting fosters financial discipline within organizations, promoting accountability and transparency in resource allocation. It enables management to make informed decisions about resource deployment, prioritize initiatives that deliver the highest returns, and allocate funds for innovation and growth. Additionally, budgets serve as a benchmark for evaluating performance across departments or business units, facilitating continuous improvement and strategic alignment with long-term goals.

FUNDING AND GRANTS

Funding is the lifeblood of organizations, providing the necessary financial resources for operations, growth, and project implementation. It encompasses various sources, including equity investments, loans, and revenue generated from operational activities. Securing adequate funding is not merely a financial imperative but a strategic necessity for sustaining operations and pursuing opportunities for expansion and development.

Types of Funding Sources

Organizations can acquire funding through different channels. Equity investments involve stakeholders buying ownership stakes in the organization, providing capital in exchange for a share in profits and governance. Loans, on the other hand, are borrowed funds that require repayment with interest over time, often secured against assets or future earnings. Revenue generated from operations represents income from selling goods or services, which can be reinvested into the organization's activities.

The Role of Grants

Grants, unlike loans, are non-repayable funds or resources disbursed by grantmakers such as government departments, corporations, foundations, or trusts. These funds are awarded to recipients, which can include nonprofit entities, educational institutions, businesses, or individuals. Grants are typically designated for specific projects or initiatives that align with the grantmaker's objectives and priorities.

Understanding the Grant Process

Securing a grant requires a systematic approach. Organizations must conduct thorough research to identify grants that match their mission and project needs. This involves understanding the criteria, eligibility requirements, and application procedures set forth by the grantmaker. Successful grant applications often demonstrate a clear alignment between the proposed project and the grantmaker's goals, highlighting the potential impact and outcomes of the project.

Grantmaker Objectives

Grantmakers distribute funds to support initiatives that advance their strategic objectives. For instance, government departments may prioritize projects that address social welfare or economic development goals. Corporations might focus on initiatives that enhance community engagement or promote sustainability. Foundations often fund projects aligned with specific causes such as education, healthcare, or environmental conservation. Understanding these priorities is crucial for effectively targeting and securing grant funding.

Impact and Sustainability

Beyond immediate financial support, grants can significantly impact an organization's sustainability and growth trajectory. Successful grant-funded projects enhance credibility, attract additional funding opportunities, and expand operational capabilities. Moreover, grants enable organizations to innovate, pilot new ideas, and address pressing societal challenges without the immediate burden of repayment.

Challenges and Considerations

While grants offer valuable financial support, the competitive nature of grant applications necessitates careful planning and strategic alignment. Organizations must invest time and resources in preparing compelling proposals that articulate their vision, capabilities, and anticipated outcomes. Additionally, managing grant funds responsibly and reporting on project progress are critical to maintaining transparency and accountability with grantmakers.

PROFITABILITY AND FINANCIAL PLANNING

Profitability and financial planning are integral components of managing a successful business, each playing a crucial role in ensuring sustainable growth and stability.

Profitability, fundamentally, measures how effectively a company converts its resources into profit. It serves as a barometer of financial health by comparing revenue generated against expenses and costs incurred over a specified period. For businesses, profitability isn't just about making money; it's about optimizing operations to achieve maximum returns. Key elements in profitability analysis include evaluating revenue growth trajectories, adept management of costs, strategic pricing strategies, and enhancing overall operational efficiency. By scrutinizing these factors, businesses can identify areas for improvement, capitalize on strengths, and mitigate weaknesses, all with the overarching goal of bolstering profit margins.

Effective financial planning complements profitability by providing a roadmap for achieving long-term financial objectives. It involves meticulous forecasting of future financial outcomes, devising comprehensive budgets, and strategizing the steps needed to attain financial stability and growth. Financial planning is a proactive approach that equips businesses to anticipate potential challenges, navigate uncertainties, and optimize cash flow management. By laying out clear financial goals and outlining the necessary actions to achieve them, businesses can make informed decisions regarding investments, resource allocation, and operational strategies.

Profitability analysis begins with assessing revenue streams and understanding the dynamics of revenue growth. This involves not only increasing sales but also diversifying income sources and improving customer retention. Concurrently, managing costs effectively is critical; businesses strive to optimize expenses without compromising quality or operational integrity. This often entails negotiating better terms with suppliers, streamlining production processes, or leveraging economies of scale.

Furthermore, pricing strategies play a pivotal role in profitability. Businesses must find a balance where pricing reflects value to customers while maximizing profit margins. This might involve market research, competitor analysis, and periodic adjustments to pricing models based on changing market conditions or customer preferences.

Operational efficiency is another cornerstone of profitability. By continually refining processes, eliminating waste, and enhancing productivity, businesses can reduce overheads and boost profitability. This might include adopting technological solutions, implementing lean management principles, or investing in employee training to improve skill sets and performance.

Financial planning, on the other hand, encompasses a broader scope beyond day-to-day operations. It involves setting specific financial goals, whether it's expanding market share, launching new products, or achieving a targeted revenue milestone. To achieve these goals, businesses develop detailed financial forecasts that project income, expenses, and cash flow over different time horizons. Budgets are then created to allocate resources effectively, ensuring that expenditures align with strategic priorities and anticipated revenue streams.

Moreover, risk management is inherent in financial planning. Businesses assess potential risks, such as economic downturns, regulatory changes, or competitive pressures, and develop contingency plans to mitigate these risks. This proactive approach not only safeguards against unforeseen disruptions but also enhances resilience and adaptability in a dynamic business environment.

RISK MANAGEMENT AND INSURANCE

Risk management is a fundamental process in financial management that encompasses the systematic identification, assessment, and prioritization of risks. Its primary objective is to coordinate efforts aimed at minimizing, monitoring, and controlling the potential impact or likelihood of adverse events. These risks can stem from diverse sources including market fluctuations, operational inefficiencies, legal liabilities, and natural calamities. Through effective risk management strategies, organizations can mitigate potential losses and safeguard their financial stability.

At its core, risk management begins with identifying potential risks relevant to an organization's operations and environment. This initial step involves a thorough examination and understanding of various risk factors that could affect the organization's objectives. Following identification, risks are assessed to determine their potential impact and likelihood of occurrence. This assessment enables prioritization, focusing resources and attention on managing risks with the highest potential impact or probability.

Once risks are identified and assessed, organizations develop strategies to manage them effectively. These strategies may include risk avoidance, risk reduction, risk sharing, or risk transfer. Risk avoidance involves taking actions to eliminate or minimize the likelihood of certain risks occurring. Risk reduction strategies aim to lessen the impact or likelihood of risks through proactive measures such as process improvements or enhanced safety protocols. Risk sharing involves distributing risk across multiple parties, while risk transfer entails transferring the financial burden of certain risks to an insurance provider.

Insurance plays a pivotal role in risk management by providing a mechanism for transferring specific risks from an organization to an insurance company. Businesses can procure insurance policies tailored to their unique needs and circumstances. These policies commonly cover areas such as property damage, liability claims, employee injuries, and business interruptions. By paying premiums, organizations transfer the financial responsibility for potential losses associated with these risks to the insurer. This transfer reduces the direct financial impact on the organization in the event of an adverse occurrence.

Moreover, insurance contributes significantly to financial stability by offering a safety net against unforeseen events that could otherwise jeopardize an organization's operations or viability. It allows businesses to allocate resources more confidently, knowing that certain risks are mitigated through insurance coverage. This confidence fosters resilience and enables organizations to focus on their core activities without undue concern about potential financial setbacks from unexpected events.

LONG-TERM ECONOMIC SUSTAINABILITY

Long-term economic sustainability is the cornerstone of organizational resilience, encapsulating an entity's capacity to uphold financial health and operational vigor over an extended period. It hinges on harmonizing economic growth with environmental stewardship and societal obligations to foster enduring viability and triumph. Realizing sustainable longevity necessitates astute strategic foresight, conscientious resource stewardship, ethical operational protocols, and adept responsiveness to evolving economic landscapes and market dynamics.

Central to long-term economic sustainability is the strategic planning framework that guides organizational decisions and actions towards enduring prosperity. This involves setting clear objectives aligned with sustainable development goals (SDGs) and integrating these goals into the core of business strategy. By embracing sustainability as a fundamental pillar, organizations can mitigate risks, enhance resilience, and capitalize on emerging opportunities that align with environmental and societal priorities.

Responsible resource management constitutes another pivotal facet of sustainable economic practices. Efficient use of resources, including raw materials, energy, and water, not only reduces costs but also minimizes environmental impact. Adoption of circular economy principles, such as recycling and waste reduction, further underscores the commitment to sustainable practices, fostering resource efficiency and resilience against resource scarcity.

Ethical business practices are integral to maintaining credibility and trust within the broader stakeholder community. Upholding integrity in dealings with employees, customers, suppliers, and the community at large strengthens the organizational reputation and enhances long-term sustainability. This includes transparency in financial reporting, adherence to regulatory requirements, and fostering a corporate culture rooted in ethical conduct and social responsibility.

Adaptability is a cornerstone of sustainable economic resilience, enabling organizations to navigate complex economic conditions and dynamic market forces. By staying attuned to shifting consumer preferences, technological advancements, and regulatory changes, businesses can innovate proactively and seize opportunities for growth while mitigating potential risks. This agility fosters long-term competitiveness and ensures organizational relevance amid evolving global trends.

Moreover, collaboration and partnerships play a pivotal role in advancing sustainable economic objectives. Engaging with stakeholders across sectors, including government bodies, non-governmental organizations (NGOs), academia, and industry peers, facilitates knowledge exchange, collective problem-solving, and the development of innovative solutions to shared sustainability challenges.

CHAPTER EIGHT

CHALLENGES AND FUTURE TRENDS

COMMON CHALLENGES IN PASTURE-RAISED PIG FARMING

Pasture-raised pig farming represents a shift away from conventional methods, prioritizing animal welfare and environmental sustainability. However, it introduces its own distinct challenges that farmers must effectively navigate.

One of the primary challenges in pasture-raised pig farming is managing the pigs' access to pasture while ensuring their health and safety. Unlike indoor confinement systems, where risks such as disease outbreaks are more contained, pasture environments expose pigs to various diseases and potential predators. Farmers must implement robust strategies, including secure fencing and regular monitoring, to protect their livestock while allowing them to benefit from outdoor grazing opportunities.

Disease management is a critical aspect of pasture-raised farming. Pigs exposed to outdoor environments are more susceptible to certain pathogens and parasites than their indoor-raised counterparts. Consequently, farmers often invest in preventive measures such as vaccination protocols tailored to outdoor conditions and strategic placement of shelters or huts that provide refuge from adverse weather and predators.

Predation poses another significant challenge in pasture-raised systems. Pigs are vulnerable to attacks from wild animals like coyotes, foxes, and even domestic dogs. Effective predator control strategies, which may include the use of guardian animals or electronic deterrents, are essential to minimize losses and ensure the safety of the herd.

Maintaining pasture quality and quantity is crucial for the sustainability of pasture-raised pig farming. Rotational grazing practices are commonly employed to prevent overgrazing and promote the regeneration of vegetation. This approach involves dividing pastures into sections and rotating pigs between them to allow for recovery periods. Farmers must possess strong land management skills to implement effective grazing schedules that optimize both animal nutrition and pasture health.

Weather variability adds another layer of complexity to pasture management. Fluctuations in rainfall and temperature can impact pasture growth and quality, influencing the availability of forage for grazing pigs. Farmers must adapt their management practices in response to changing weather patterns, potentially adjusting grazing rotations or supplementing with stored forage during periods of scarcity.

In addition to these operational challenges, pasture-raised pig farming requires a commitment to environmental stewardship. Farmers often integrate sustainable practices such as composting manure to enrich soil fertility, planting cover crops to prevent erosion, and managing water resources responsibly to minimize environmental impact.

Despite these challenges, pasture-raised pig farming offers significant advantages, including improved animal welfare, enhanced product quality, and consumer demand for ethically produced meat products. Successful navigation of these challenges requires a combination of practical experience, ongoing education, and a willingness to adapt to changing conditions in both agricultural and environmental landscapes.

Overall, while pasture-raised pig farming presents unique obstacles compared to conventional methods, proactive management strategies can mitigate risks and support sustainable production practices that benefit both animals and the environment. By addressing challenges such as disease management, predator control, pasture quality, and environmental sustainability, farmers can maximize the potential of pasture-raised systems while meeting the growing demand for responsibly sourced pork products.

INNOVATIONS AND TECHNOLOGICAL ADVANCES

In addressing the challenges of pasture-raised pig farming, significant innovations and technological advances have emerged to bolster efficiency, sustainability, and animal welfare. These advancements span various domains, from precision monitoring using GPS and sensor technologies to genetic enhancements and sustainable feeding practices.

One of the key technological innovations transforming pasture-raised pig farming is the integration of GPS tracking and sensor technology. These tools enable farmers to remotely monitor pig movements and health parameters such as temperature and activity levels. By collecting real-time data, farmers can swiftly detect any anomalies, facilitating early intervention in case of illness or distress. This proactive monitoring not only enhances animal welfare but also optimizes management practices by providing insights into herd behavior and health trends.

Additionally, automated feeding systems have revolutionized how pigs are fed in outdoor environments. These systems dispense precise quantities of feed at scheduled intervals, ensuring consistent nutrition while reducing labor costs and minimizing environmental impact. Coupled with advancements in solar-powered infrastructure, which powers these automated systems sustainably, farmers can achieve greater operational efficiency and reduce their carbon footprint.

Genetic advancements play a pivotal role in improving the performance and resilience of pasture-raised pigs. Selective breeding programs focus on traits such as disease resistance and efficient pasture utilization. By breeding pigs that thrive in outdoor settings and exhibit robust health, farmers can mitigate disease risks and enhance overall productivity. These genetic improvements not only benefit the pigs but also contribute to sustainable farming practices by reducing the need for antibiotics and other interventions.

Furthermore, ongoing research into alternative feeds and supplements aims to enhance the nutritional profiles of pasture-raised pigs. By diversifying feed sources and incorporating supplements that complement outdoor foraging, farmers can ensure balanced diets that meet the pigs' nutritional requirements. This approach reduces dependency on traditional feed sources, thereby promoting sustainability and cost-effectiveness in pig farming operations.

Innovative technologies are also facilitating data-driven decision-making in pasture-raised pig farming. By analyzing data collected from GPS trackers, sensors, and automated systems, farmers can optimize feeding strategies, manage pasture utilization more effectively, and improve overall farm productivity. This integration of technology not only streamlines farm operations but also empowers farmers with actionable insights to make informed decisions that enhance both economic and environmental sustainability.

CONSUMER TRENDS AND MARKET DEMANDS

Consumer trends in the food industry are undergoing a notable shift towards sustainably produced foods, particularly pasture-raised pork. This changing consumer preference is largely motivated by growing concerns over animal welfare, environmental sustainability, and food safety. As a result, there has been a noticeable increase in demand for pork products that are sourced from animals raised on pasture.

The shift towards pasture-raised pork can be attributed to several key factors. Firstly, consumers are increasingly aware of and concerned about the welfare of animals raised for food. Pasture-raised practices are perceived as more humane, allowing pigs to roam outdoors, exhibit natural behaviors, and have access to natural forage. This contrasts with conventional intensive farming methods where animals are often confined indoors in close quarters.

Environmental sustainability is another significant driver of this trend. Pasture-raised pork production is seen as more environmentally friendly compared to conventional methods. Grazing animals on pasture can help improve soil health, reduce reliance on chemical fertilizers, and mitigate environmental impacts associated with concentrated animal feeding operations (CAFOs).

Furthermore, there is a heightened awareness of food safety issues among consumers. Pasture-raised pork is often perceived as safer and healthier due to the animals' natural diet and lower exposure to antibiotics and other medications commonly used in intensive farming.

The willingness of consumers to pay premiums for pasture-raised pork underscores the economic incentives for farmers to adopt these production practices. This premium pricing reflects consumers' valuation of higher welfare standards, environmental stewardship, and perceived health benefits associated with pasture-raised products. For farmers, transitioning to pasture-raised systems not only meets consumer demand but also potentially enhances their profitability through differentiated products in a competitive market.

To help consumers navigate these choices, certification programs and labeling schemes play a crucial role. Certifications such as organic and pasture-raised provide consumers with assurance that specific production standards have been met. Organic certification, for instance, guarantees that animals have been raised according to strict organic standards, which include access to pasture and restrictions on synthetic inputs.

Similarly, pasture-raised certifications validate that animals have been raised primarily on pasture and have had ample opportunity for outdoor exercise and natural foraging. These labels not only inform consumers about the production methods but also serve as valuable marketing tools for producers looking to differentiate their products in a crowded marketplace.

ADAPTING TO CLIMATE CHANGE

Adapting to climate change is becoming increasingly crucial for pasture-raised pig farming, presenting both challenges and opportunities. With the unpredictable nature of extreme weather events and shifting precipitation patterns, the resilience of pasture productivity and animal health is at stake. Farmers are compelled to proactively adopt adaptive measures to sustain their operations in the face of these challenges.

One of the primary concerns for pasture-based pig farming is the impact of extreme weather events. Floods, droughts, and heatwaves can disrupt pasture growth cycles and stress animal health. To mitigate these risks, farmers are exploring various strategies. Implementing efficient water management systems helps to conserve water during dry periods and mitigate flood damage during heavy rains. By optimizing irrigation techniques and storage facilities, farmers can better regulate water availability for their pigs and pasturelands.

Furthermore, selecting drought-resistant pasture species is crucial. These species can endure prolonged dry spells and require less water, thus ensuring continuous forage availability even in arid conditions. Farmers are increasingly diversifying their pasture mix to include such resilient species, thereby safeguarding against the adverse effects of erratic precipitation patterns.

Improving shelter designs is another critical adaptation strategy. Enhanced shelters provide refuge from extreme weather, offering pigs protection from harsh elements such as storms and excessive heat. By integrating climate-responsive design features, such as natural ventilation and thermal insulation, farmers can maintain optimal conditions within shelters regardless of external weather variations.

Beyond adaptation, pasture-raised pig farming holds potential as a contributor to climate change mitigation through carbon sequestration. Rotational grazing practices, where pigs are moved between different pasture areas, promote healthier soils that act as carbon sinks. These soils trap and store carbon dioxide from the atmosphere, offsetting greenhouse gas emissions associated with agricultural activities. The integration of cover cropping and minimal tillage practices further enhances soil health and carbon storage capacity, making pasture-raised systems a sustainable choice amidst climate challenges.

THE FUTURE OF SUSTAINABLE PIG FARMING

Looking forward, the trajectory of pasture-raised pig farming rests on ongoing innovation, adaptable strategies, and active consumer involvement. Technological advancements are poised to significantly enhance efficiency and sustainability benchmarks, marking a critical evolution in the industry. The collaboration between farmers, scientists, and policymakers will prove indispensable in tackling emerging obstacles and leveraging new prospects.

In the coming years, technological innovations are expected to revolutionize sustainable pig farming. These advancements may encompass precision farming techniques, such as AI-driven monitoring systems that optimize feed efficiency and health management. Additionally, developments in renewable energy integration and waste management solutions promise to further mitigate environmental impacts while enhancing operational sustainability. By embracing these technologies, farmers can not only streamline their processes but also bolster their environmental stewardship credentials.

Collaborative research initiatives will play a pivotal role in shaping the future landscape of pig farming. By fostering partnerships between agricultural experts, researchers, and policy makers, the industry can proactively address challenges such as disease management, resource allocation, and regulatory compliance. This collective approach not only enhances knowledge sharing but also promotes the adoption of best practices across the sector.

Education and outreach efforts are essential in fostering consumer awareness and support for sustainable farming practices. By transparently communicating the benefits of pasture-raised systems—such as improved animal welfare, reduced environmental footprint, and higher food quality—farmers can cultivate a loyal customer base. Initiatives aimed at promoting these benefits through marketing campaigns, farm tours, and community engagements are crucial in building consumer trust and securing market opportunities.

Consumer engagement plays a pivotal role in driving demand for sustainably produced pork products. As awareness grows about the environmental and ethical considerations associated with food production, there is an increasing preference for products that align with these values. Farmers who prioritize transparency in their practices and actively engage with consumers through social media, farmer's markets, and educational workshops are well-positioned to capitalize on this trend. By establishing direct relationships with consumers, farmers can not only secure market access but also receive valuable feedback that informs their farming practices.

Looking ahead, the future of pasture-raised pig farming hinges on continuous adaptation and innovation. By leveraging technological advancements, fostering collaborative research efforts, and engaging consumers through education and outreach, the industry can navigate challenges and capitalize on opportunities for sustainable growth. This proactive approach not only ensures the viability of pasture-raised systems but also reinforces their role in meeting the evolving expectations of consumers and society at large.

THE END